BOXER BOOKS Ltd. and the distinctive Boxer Books logo are trademarks of Union Square & Co., LLC.

Union Square & Co., LLC, is a subsidiary of Sterling Publishing Co., Inc.

Illustrations copyright © 2009 Britta Teckentrup
Text copyright © 2009 Boxer Books Limited

All rights reserved. No part of this publication may be reproduced, stored in a retrieval system, or transmitted in any form or by any means (including electronic, mechanical, photocopying, recording, or otherwise) without prior written permission from the publisher.

This hardback edition first published in 2023
by Boxer Books Limited.

Originally published as Big Noisy Book of Busy Animals 2009

ISBN 978-1-914912-53-5

A catalogue record of this book is available from the British Library.

For information about custom editions, special sales, and premium purchases,
please contact specialsales@unionsquareandco.com.

Printed in China.

2 4 6 8 10 9 7 5 3 1

12/22

unionsquareandco.com

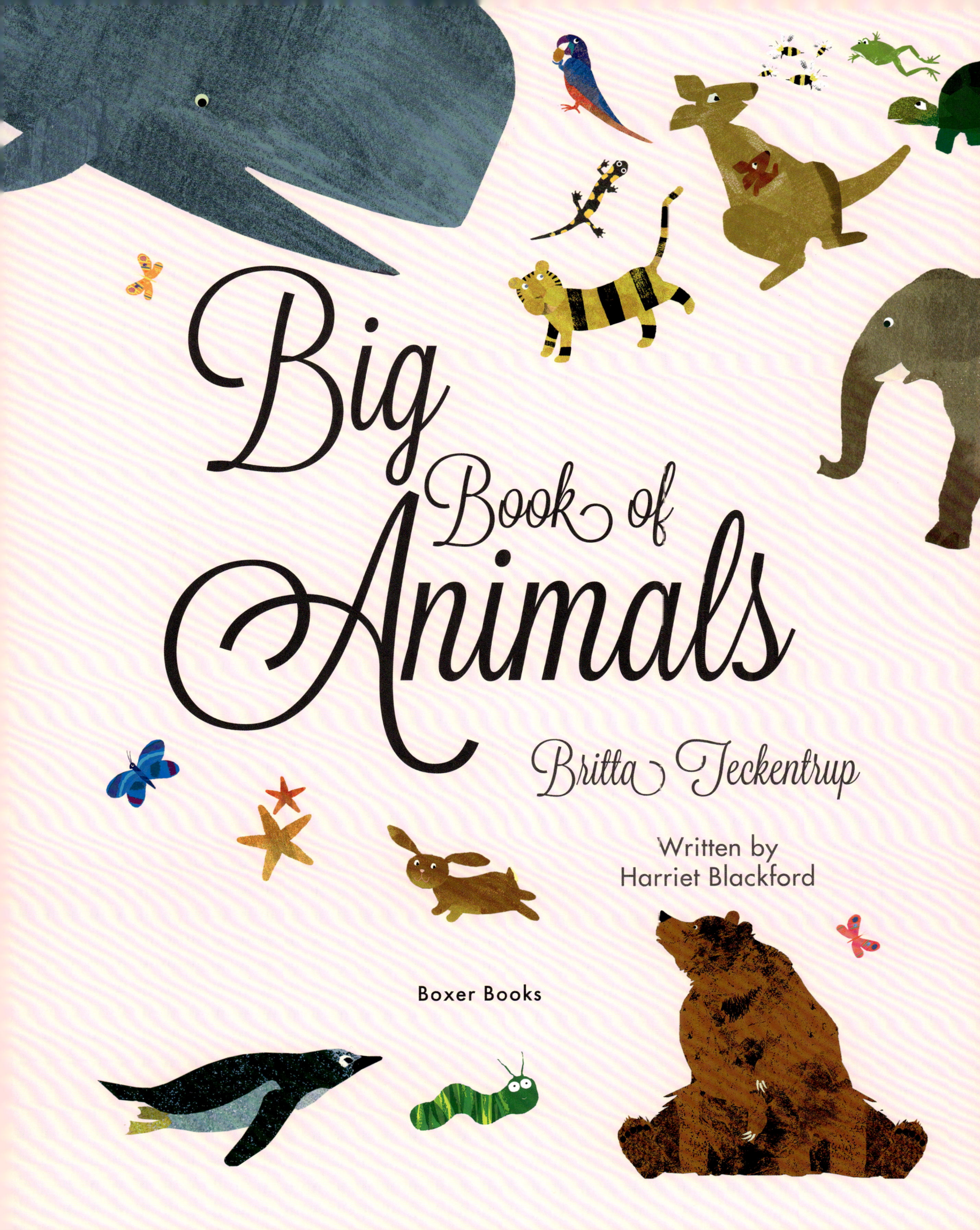

Animals

WHAT IS AN ANIMAL? 8–9

ANIMALS ALL OVER THE WORLD 10–11

ANIMAL SHAPES 12–13

COLOURS AND PATTERNS 14–15

ANIMALS WITH BONES 16–17

ANIMALS WITHOUT BONES 18–19

HOW DO ANIMALS MOVE? 20–21

WHAT DO ANIMALS EAT? 22–23

POO FACTS 24–25

ANIMAL HOMES 26–27

ANIMAL BABIES 28–29

GROWING UP 30–31

ANIMAL MANNERS 32–33

BIGGEST AND SMALLEST 34–35

BIZARRE AND WEIRD 36–37

Animals need to breathe air.

Some even breathe underwater.

Animals have babies.

Animals start small and grow bigger.

Animals poo and wee to get rid of waste from their bodies.

You are an animal.

ANIMALS ALL OVER THE WORLD

Animals live all over the world.

Eagles soar high in the sky.

Snow leopards pounce high up in the mountains.

Worms burrow through earth.

Ducks paddle in ponds.

Cockroaches scurry in towns and cities.

Alligators swim in swamps.

Camels walk across hot, dry deserts.

Octopuses hide deep down in the sea.

Polar bears live in cold, icy places.

Rabbits hop around the fields.

Howler monkeys call out in the hot, steamy jungle.

ANIMAL SHAPES

The starfish looks like a star with five arms.

Spiders have lots of legs. Count them.

1 2 3 4 5 6 7 8

Snakes do not have arms or legs!

Hammerhead sharks have strange-shaped faces.

Jellyfish look like blobs of jelly and have no face at all.

The flying squirrel makes a square in the air as it glides from tree to tree.

COLOURS AND PATTERNS

The ladybird is red and spotted.

It is difficult to see one zebra in a crowd, because they all have stripes.

The giraffe has a pattern of patches on its body.

Stripes make it hard to see the tiger in the long grass.

The panda has patches around its eyes.

Bright yellow cichlid fish cruise tropical waters.

Velvet-black bats swoop through the night air.

Vivid green katydids hide in the long grass.

Red-bottomed baboons stand out in a crowd.

A peacock has eye patterns on its tail feathers. What a show-off!

Iridescent blue poison arrow frogs croak in the jungle.

Birds come in all colours of the rainbow!

A stick insect is brownish green, just like a stick. How many stick insects can you see?

ANIMALS WITH BONES
Some animals have bones inside their body. This is called a skeleton. A skeleton gives the animal shape and helps it move.

The furry ones are called mammals, like this gorilla feeding its baby.

Animals with feathers are called birds. Most birds can fly.

You are a mammal too, but your fur is called hair.

Penguins cannot fly in the air, but they can swim through the water, just as if they were in the sky.

The albatross has the longest wings of any bird.

Lumbering tortoises have hard shells.

Slithery snakes have dry, shiny, scaly skin.

Scaly animals are called reptiles.

Toothy crocodiles have thick skin.

Quick-moving lizards come in many colours.

Frogs, toads, newts and salamanders are amphibians. They live in the water and on the land.

Fish can only live in water.

ANIMALS WITHOUT BONES
Some animals do not have bones. Instead of bones some have hard, crusty skin and lots of legs.

Scuttling beetles.

Fluttering butterflies.

Web-spinning spiders.

Scorpions with a stinger in their tail.

Crabs walk sideways.

Woodlice under the woodpile.

Centipedes and millipedes have lots and lots and lots of legs!

Instead of bones, or hard crusty skin, some animals are squishy.

Earthworms burrow through the earth.

Jellyfish, squid and beautiful sea slugs float underwater.

Molluscs are squishy and make shells to live in.

Snails are molluscs, and so are the animals that make the seashells that you find on the beach.

HOW DO ANIMALS MOVE?

Frogs leap high to reach the pond.

Fleas jump 50 times their body length.

Horses run on their hooves.

Birds, bats and insects fly.

Fish swim using their fins and tails.

Foxes stalk and pounce and rabbits hop.

Dolphins use fins and tails too, and leap right out of the water.

Monkeys climb trees. Some can hang by their tails.

Octopuses use jet propulsion to move along.

Slugs and snails slither along the ground, leaving a slimy trail.

Sideways-scuttling crabs race across the beach.

Snakes grip the ground with their ribs to move forward.

Sea anemones wave their tentacles but shrink to a blob if frightened.

But nothing hurries a sloth!

Sloths move VERY slowly!

WHAT DO ANIMALS EAT?
**Some animals only eat plants.
Some animals eat other animals.**

Koalas sit up in trees, chewing on leaves.

Cows and sheep spend all day grazing on grass.

Chimpanzees love fruit.

Insects all over the world are munching the plants around them.

Parrots crack nuts with their strong bills.

Snakes swallow their prey whole.

Frogs use their long, sticky tongues to catch insects.

Wolves hunt other animals in packs.

Spiders trap flies in their webs.

Bears eat lots of different things — fish, berries, honey and meat.

What do you like to eat?

Jellyfish trap fish in their tentacles.

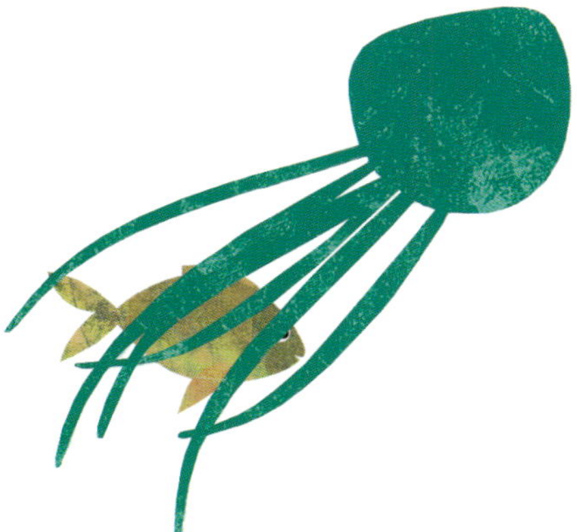

POO FACTS

Poo is the waste left over after an animal has eaten its food.

Elephants make lots of poo. Their poo is called dung.

Dung beetles roll it away to lay their eggs in it, and they eat it!

A week's worth of elephant dung weighs as much as 70 children!

Whale poo sinks to the ocean floor.

Rabbits eat their poo to get all the goodness from the grass they have eaten.

Earthworms eat earth that has tiny bits of plants in it. They poo out a curly cast.

Birds have whitish poo — unless they have been eating blackberries!

Birds do not wee.

Insects leave a trail of very tiny poo.

Hippos wag their tails when they are pooing. They leave a trail so they can find their way back to the river.

Cats are tidy and like to bury their poo.

ANIMAL HOMES
Some animals build homes to live in.

Beavers make dams by gnawing down trees and blocking streams with them.

They build their home, called a lodge, with the entrance underwater.

Termites are tiny but build huge mounds to live in.

Birds make nests to lay their eggs in.

Weaverbirds build fantastic nests!

Rabbits dig underground burrows called warrens.

Gorillas make themselves a new nest to sleep in every night.

Some bats spend the daytime hanging upside down in caves.

Wasps make beautiful nests from tiny bits of wood and saliva.

Snails and tortoises carry their homes with them.

When hermit crabs outgrow their shell, they find an empty one and wriggle into it as quickly as possible.

Some animals live with people in their homes.

Some animals are not always welcome!

ANIMAL BABIES

Some baby animals look just like their parents, only smaller.

Look at all these piglets feeding from their mother.

This mother spider has hundreds of babies and they all look just like her.

Baby whales are helped to the surface to take their first breath after being born.

Some baby animals look very different from their parents.

Caterpillars are baby butterflies or moths!

Tadpoles look nothing like their froggy mum.

Some babies need a lot of care.

Human babies have to be carried everywhere until they learn to walk! Even then, they still need care for many years.

Chicks are brought food by their parents.

Baby stick insects start munching leaves right away.

Some babies look after themselves as soon as they are born.

Newly hatched turtles run straight to the sea.

29

GROWING UP

Some babies have to change a lot before they are grown up, whereas others have a lot to learn.

Tadpoles stay in the pond until their legs grow and their tail shrinks.

A caterpillar eats and eats.

It makes a hard skin, a cocoon, to sleep inside while its body slowly changes into a butterfly!

Tiger cubs learn to hunt by copying their mother and playing with their brothers and sisters.

Chimpanzee babies copy the grown-ups to find out what is good to eat and how to use sticks to get insects out of holes.

Children have a lot to learn before they grow up to be adults. They spend many years in school.

Some baby animals, like grasshoppers and snakes, have to shed their skin to grow bigger.

Baby birds exercise their wings to get ready for flight.

Lots of baby animals play to help strengthen their muscles and to learn how to live with one another.

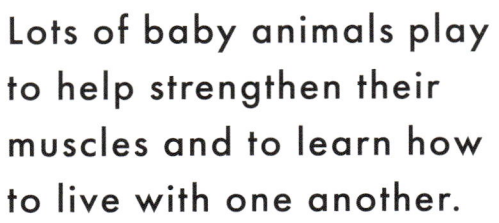

ANIMAL MANNERS

We learn manners so we can all live together peacefully.

We shake hands and smile to say hello!

Some people bow to each other.

Dogs sniff each other's bottoms to find out how old or healthy another dog is.

Wagging tails mean happy dogs.

But when a cat swishes its tail it is not happy!

Angry horses and zebras flatten their ears.

Birds sing to say, "This is my space".

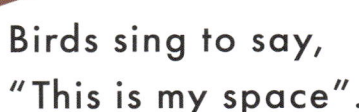

Other animals, like tigers, dogs and bears, say, "This is my space" by weeing onto rocks and trees. This is called marking your territory.

Monkeys groom each other to make and keep friends.

Male spiders are much smaller than female spiders. To save themselves from being eaten, they give a gift of food when they mate.

BIGGEST AND SMALLEST

The biggest animal in the world is the blue whale. It is as long as 39 people lying down head to toe.

Tiny creatures in the sea called zooplankton are so small you have to use a microscope to see them.

The smallest mammal is the bumblebee bat. This is its real size.

The whale shark would need a fish tank as big as a public swimming pool.

The biggest eye in the world belongs to the colossal squid. It is more than 60 centimetres wide.

Giraffes are the tallest land animals. Six chimpanzees would have to stand on each other's shoulders to look a giraffe in the eye.

The smallest bird is the bee hummingbird at 5 centimetres.

The biggest bird is the ostrich. One of its eggs would make an omelette for 10 people!

The biggest spider is the goliath bird-eating spider. It is as big as your dinner plate.

African elephants are the biggest and heaviest animals on land. One elephant weighs more than four million mice!

BIZARRE AND WEIRD
Animals have some ways of living that might look weird or seem strange.

Wallace's flying frog stretches out its big webbed feet to glide.

Flying dragon lizards can glide from tree to tree by stretching out the skin on their sides.

The opossum plays dead if it is frightened. It falls over with its mouth open and gives off a smell of rotting meat!

The male hooded seal can blow out the lining of its nose to make a big bubble if it is angry or if it wants to show off.

The mantis shrimp kills its food with a super-fast punch from one of its legs.

Fireflies can make light in their bodies to attract a mate.

The pink fairy armadillo can bury itself completely in seconds if it is frightened!

Beware the skunk! It sprays out the worst smell ever.

Archerfish spit at insects and knock them down into the water to eat.

The duck-billed platypus is a mammal that lays eggs and has a duck-like bill!